THOMAS CRANE PUBLIC LIBRARY
QUINCY MA

CITY APPROPRIATION

SPACE FRONTIERS

The Universe

Helen Whittaker

A+

This edition first published in 2011 in the United States of America by Smart Apple Media.

All rights reserved. No part of this book may be reproduced in any form or by any means without written permission from the Publisher.

Smart Apple Media
P.O. Box 3263
Mankato, MN, 56002

First published in 2010 by
MACMILLAN EDUCATION AUSTRALIA PTY LTD
15–19 Claremont Street, South Yarra 3141

Visit our website at www.macmillan.com.au or go directly to www.macmillanlibrary.com.au

Associated companies and representatives throughout the world.

Copyright © Macmillan Publishers Australia 2010

Library of Congress Cataloging-in-Publication Data

Whittaker, Helen, 1965-
 The universe / Helen Whittaker.
   p. cm. — (Space frontiers)
 Includes index.
 ISBN 978-1-59920-576-2 (lib. bdg.)
 1. Galaxies—Juvenile literature. 2. Stars—Evolution—Juvenile literature.
 3. Expanding universe—Juvenile literature. I. Title.
 QB857.3.W47 2011
 523.1'12—dc22

                                    2009038482

Edited by Laura Jeanne Gobal
Text and cover design by Cristina Neri, Canary Graphic Design
Page layout by Cristina Neri, Canary Graphic Design
Photo research by Brendan and Debbie Gallagher
Illustrations by Alan Laver, except page 18 by Melissa Webb

Manufactured in China by Macmillan Production (Asia) Ltd.
Kwun Tong, Kowloon, Hong Kong
Supplier Code: CP December 2009

Acknowledgments
The author and the publisher are grateful to the following for permission to reproduce copyright material:

Front cover photos of artwork of a spherical universe, courtesy of Photolibrary/Mehau Kulyk/Science Photo Library; blue nebula background © sololos/iStockphoto.

Photographs courtesy of:
Chandra X-ray Observatory Center, **23**; © Roger Ressmeyer/Corbis, **27**; ESA & NASA/E. Olszewski (U. Arizona) HST, **9**; Hulton Archive/Getty Images, **5**; © pederk/iStockphoto, **6** (main); © sololos/iStockphoto **6–7** (background); © susaro/iStockphoto, **7** (main); NASA/CXC/M.Weiss, **8**, **17**; NASA, ESA and G. Bacon (STScI), **13**, **29**; NASA, ESA, CXC, JPL-Caltech, J. Hester and A. Loll (Arizona State Univ.), R. Gehrz (Univ. Minn.) and STScI, **3**, **15**; NASA, ESA, M. J. Jee and H. Ford (Johns Hopkins University), **19**; NASA, ESA, M. Robberto (Space Telescope Science Institute/ESA) and the Hubble Space Telescope Orion Treasury Project Team, **10** (bottom right); NASA, ESA, N. Smith (University of California, Berkeley) and The Hubble Heritage Team (STScI/AURA), for Hubble image, and N. Smith (University of California, Berkeley) and NOAO/AURA/NSF for CTIO Image, **10** (top); NASA and Ann Feild (STScI), **26**; NASA and The Hubble Heritage Team (AURA/STScI), **12** (bottom); NASA/JPL-Caltech/R. Hurt (SSC-Caltech), **21**, back cover; NASA/JPL-Caltech/Howard McCallon, **10** (bottom left); NASA/JSC, Eugene Cernan, **4**; NASA/MSFC, **14**; NASA/MSFC/ESA, H. E. Bond (STScI) and The Hubble Heritage Team (STScI/AURA), **12** (top and center); NASA/WMAP Science Team, **25**; Photolibrary/Landolfi Larry, **20**; Photolibrary/Lynette Cook/SPL, **28**; Photolibrary/Detlev Van Ravenswaay/SPL, **30**; Aurore Simonnet, Sonoma State University, **22**.

Images used in design and background on each page © prokhorov/iStockphoto, Soubrette/iStockphoto.

While every care has been taken to trace and acknowledge copyright, the publisher tenders their apologies for any accidental infringement where copyright has proved untraceable. Where the attempt has been unsuccessful, the publisher welcomes information that would redress the situation..

CONTENTS

Space Frontiers   4
The Universe   5
A Timeline of the Universe   6

Stars
What Is a Star?   8
The Life Cycle of a Star   10
Variable Stars   12
Neutron Stars   14
Black Holes   16

Galaxies
What Is a Galaxy?   18
The Milky Way   20
Active Galaxies   22

The Universe
The Expanding Universe   24
The Birth of a Theory   26
Looking for Life   28

The Future of the Universe   30
Glossary   31
Index   32

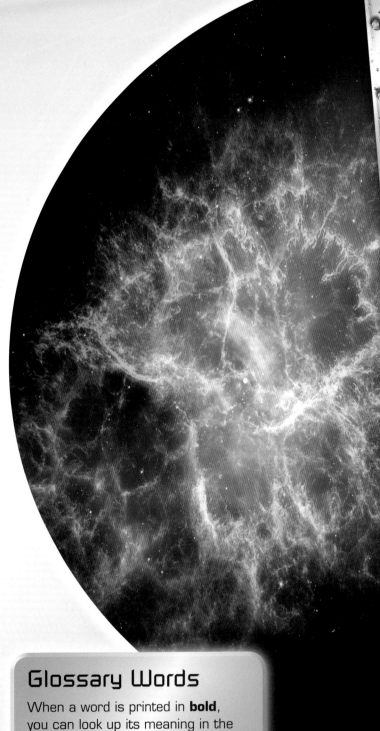

Glossary Words
When a word is printed in **bold**, you can look up its meaning in the Glossary on page 31.

SPACE FRONTIERS

A frontier is an area that is only just starting to be discovered. Humans have now explored almost the whole planet, so there are very few frontiers left on Earth. However, there is another frontier for us to explore and it is bigger than we can possibly imagine—space.

Where Is Space?

Space begins where Earth's **atmosphere** ends. The atmosphere thins out gradually, so there is no clear boundary marking where space begins. However, most scientists define space as beginning at an altitude of 62 miles (100 km). Space extends to the very edge of the universe. Scientists do not know where the universe ends, so no one knows how big space is.

Exploring Space

Humans began exploring space just by looking at the night sky. The invention of the telescope in the 1600s and improvements in its design have allowed us to see more of the universe. Since the 1950s, there has been another way to explore space—spaceflight. Through spaceflight, humans have **orbited** Earth, visited the Moon, and sent space probes, or small unmanned spacecraft, to explore our **solar system**.

▲ Spaceflight is one way of exploring the frontier of space. Astronaut Harrison Schmitt collects Moon rocks during the Apollo 17 mission in December 1972.

THE UNIVERSE

The universe consists of everything that exists. It is made up of time, space, matter, and energy. Our understanding of the universe has changed a lot over the past 500 years.

How Did We First View the Universe?

For most of human history, astronomers thought Earth was at the center of the universe, with the Sun, the Moon, and the planets moving around it, each attached to the surface of one hollow, transparent ball called a celestial sphere. The stars were attached to a single, rotating outer sphere. The universe was believed to be endless and unchanging. This theory became known as the geocentric model.

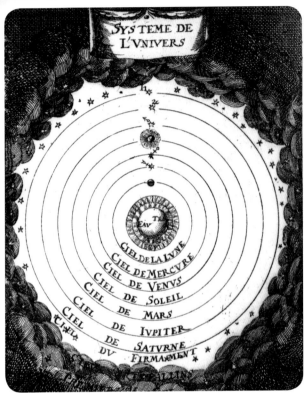

The geocentric model of the universe proposed that Earth was at the center. Surrounding it were the Moon, Mercury, Venus, the Sun, Mars, Jupiter, Saturn, and the stars.

Why Did Our View of the Universe Change?

In the 200s BC, the Greek philosopher Aristarchus suggested the Sun was at the center of the universe and that Earth and all the other planets orbited it. This theory, known as heliocentrism, was largely ignored. Then in 1543, the Polish astronomer Nicolaus Copernicus used heliocentrism to explain the apparent backward motion of Mars, Jupiter, and Saturn. The theory gradually gained support from other astronomers and became the most widely accepted model of the universe.

How Do We Now View the Universe?

We now know that Earth is definitely not at the center of the universe. Earth and all the planets in our solar system revolve around the Sun in a **galaxy** known as the Milky Way. The Sun is one of more than 200 billion stars in this galaxy, which is one of more than 100 billion galaxies. The universe is neither endless nor unchanging. In fact, it is expanding incredibly fast as a result of the explosion that created it about 13.7 billion years ago. This explosion is known as the **big bang**.

A TIMELINE OF THE UNIVERSE

The timeline below shows some of the most important events in the history of the universe.

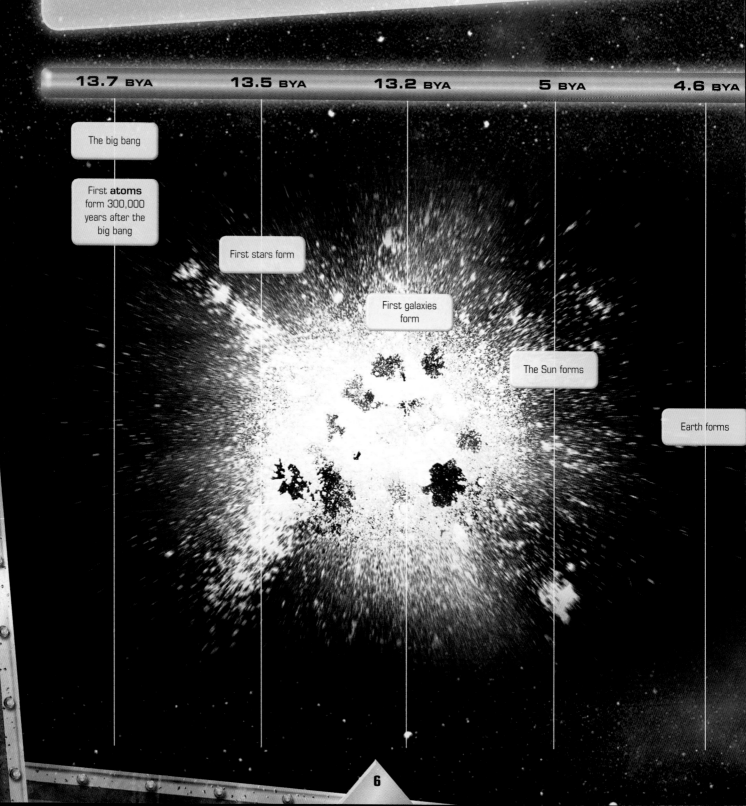

13.7 BYA — The big bang

First **atoms** form 300,000 years after the big bang

13.5 BYA — First stars form

13.2 BYA — First galaxies form

5 BYA — The Sun forms

4.6 BYA — Earth forms

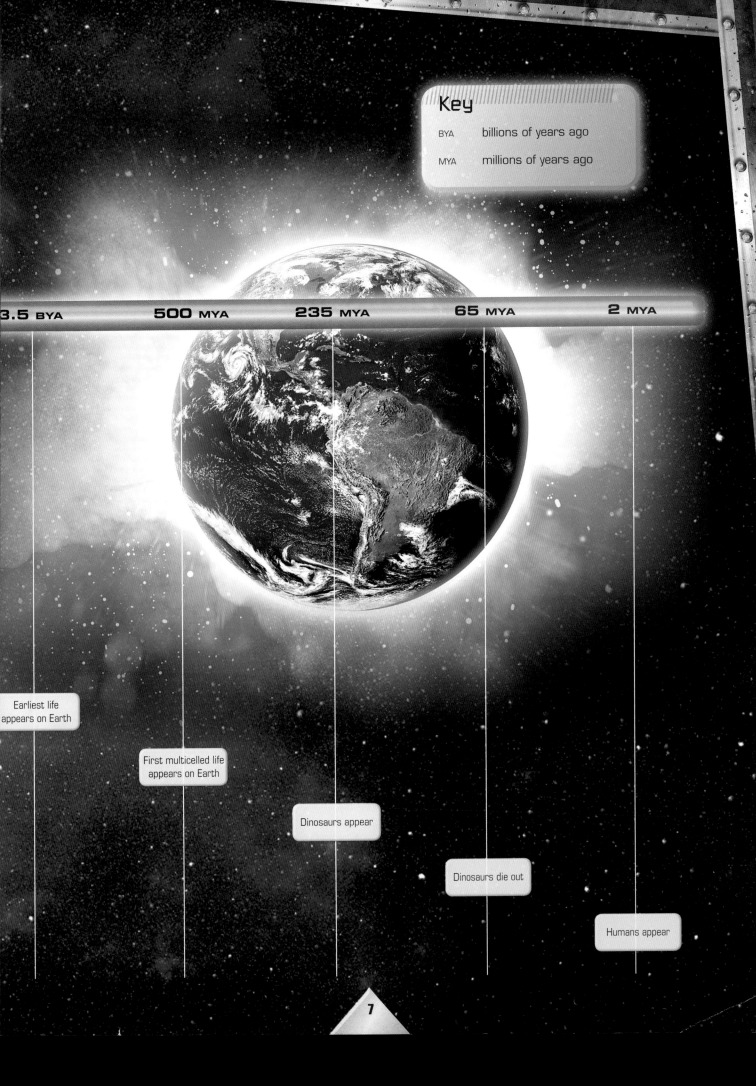

What is a Star?

Stars

A star is a massive, glowing ball of **plasma**, which is held together in a round shape by its own **gravity**.

Why Do Stars Shine?

Stars shine because they produce huge amounts of energy in their core through **nuclear fusion**. The energy released takes the form of visible light, heat, and other types of **electromagnetic radiation**.

▼ **This labeled illustration shows the structure of a Sun-like star. Nuclear fusion takes place in the core.**

Our Star, the Sun

The Sun is a star. It has a diameter of at least 864,327 miles (1.39 million km), which is about 108 times wider than Earth. The Sun is composed of 92.1 percent hydrogen and 7.8 percent helium. Small quantities of other **elements** are also present. The Sun has a surface temperature of about 10,000 °F (5,500 °C) but temperatures at its core reach more than 27 million °F (15 million °C).

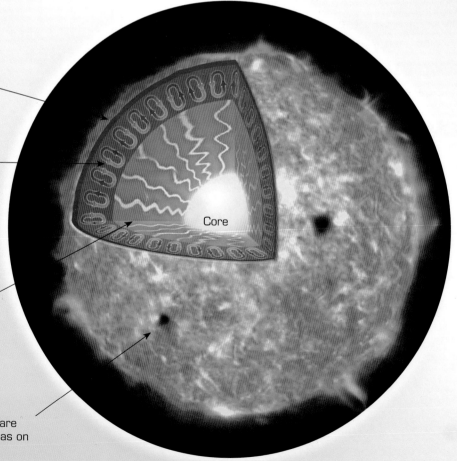

Photosphere This is the visible surface of a star.

Convection zone Matter is packed less tightly and is able to move around. In this zone, heat from the core is transferred outward by the process of convection (the way heat moves through currents of plasma).

Radiation zone Matter is packed together very tightly, and heat from the core is transferred outward by the process of radiation (the direct transfer of energy from one particle to another).

Sunspots These are darker, cooler areas on the photosphere.

▲ The universe contains stars of all sizes, colors, and magnitudes. This image shows a cluster of stars about **200,000 light-years** away, in a galaxy known as the Small Magellanic Cloud.

Classifying Stars

Stars can be classified in three main ways. One way astronomers classify, or group, stars is by their size. Stars can also be classified according to color and magnitude.

Size

The Sun belongs to the category of smallest stars. It is a dwarf star. Giant stars are between 10 and 100 times as big as the Sun, while supergiants can be up to 1,000 times as big.

Color

A star's color depends on how hot it is. Red stars are the coolest, with surface temperatures of 6,332°F (3,500°C), while blue stars are the hottest, at 90,032°F (50,000°C).

Did You Know?

The Sun is a main sequence star. Main sequence stars are stars in the adult phase of their life cycle, during which time nuclear fusion in their core converts hydrogen into helium.

Magnitude

A star's brightness is measured on a scale known as magnitude. The faintest stars have a high magnitude, while the brightest have a low magnitude. Sirius, the brightest star in the sky, has a negative magnitude of -1.46. Stars with magnitudes of up to six are visible to the naked eye.

THE LIFE CYCLE OF A STAR

Although stars seem unchanging to us, they have a life cycle just as living things do. The only difference is that stars live for billions of years.

The Birth of a Star

Stars form in a huge cloud of gas and dust called a nebula. As gravity pulls matter in the nebula together, clumps are formed, which heat up. Each clump forms a spinning disc called a protostar. Eventually, the center of the protostar gets so hot that nuclear fusion begins. The solar wind, a stream of charged particles ejected from the upper atmosphere of the forming star, blows the remaining dust away, and a star is born.

New stars are born all the time. The Carina Nebula, shown here, is about 7,500 light-years away. New stars are forming inside the towering clouds of gas and dust.

Find a Star-Forming Nebula

The Orion Nebula is a region of intense star formation that is visible to the naked eye. It is the brightest object in what is known as Orion's sword, which is a small cluster of three stars below Orion's belt (in the Northern Hemisphere) or above the belt (in the Southern Hemisphere). The Orion Nebula is the middle "star" in the sword.

This is part of the **constellation** of Orion. Orion's belt is the group of three stars in the middle. The Orion Nebula is the bright purple area in the section below it, known as Orion's sword.

The Life Cycle of Dwarf Stars

A star's life cycle depends on its mass. Dwarf stars, like the Sun, are the most common type of star. They shine for a few billion years as a main sequence star. Then they expand to become **red giants** before shedding their outer layers and becoming **planetary nebulae**. Eventually, they end their days as small, cold, **white dwarfs**.

The Life Cycle of Massive Stars

Massive stars have a more dramatic life cycle. When hydrogen in their core runs out, they expand to become supergiants, before tearing themselves apart in an enormous explosion called a supernova. After the supernova, all that is left of the star is an extremely hot and dense **neutron star** or, if the remains are massive enough, a black hole.

> ### Did You Know?
> The last supernova that occurred in the Milky Way was observed in 1604. It was named Kepler's Supernova after Johannes Kepler, the German astronomer who was the first to observe it. The supernova was visible to the naked eye for about 18 months and, at its brightest, was brighter than Jupiter.

▼ A star's life cycle depends on its mass. Massive stars die dramatically in supernova explosions. Less massive stars fade away gradually.

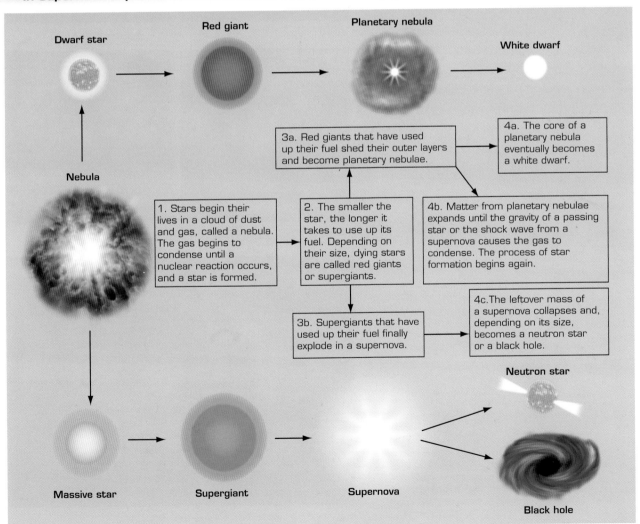

1. Stars begin their lives in a cloud of dust and gas, called a nebula. The gas begins to condense until a nuclear reaction occurs, and a star is formed.
2. The smaller the star, the longer it takes to use up its fuel. Depending on their size, dying stars are called red giants or supergiants.
3a. Red giants that have used up their fuel shed their outer layers and become planetary nebulae.
3b. Supergiants that have used up their fuel finally explode in a supernova.
4a. The core of a planetary nebula eventually becomes a white dwarf.
4b. Matter from planetary nebulae expands until the gravity of a passing star or the shock wave from a supernova causes the gas to condense. The process of star formation begins again.
4c. The leftover mass of a supernova collapses and, depending on its size, becomes a neutron star or a black hole.

VARIABLE STARS

A variable star is one whose brightness varies, or changes. Some stars alternate between bright and dim. Such stars are called intrinsic variables. Other stars, called extrinsic variables, only appear to change because something else blocks their light.

Intrinsic Variables

Intrinsic variables are stars whose brightness varies because of changes in the star itself. Some intrinsic variables have a cycle of growing and shrinking. Others go through periods of violent storms and are much brighter when the storms are raging. Some intrinsic variables have enormous sunspots, which are cool, dark areas. When these sunspots are pointing toward Earth, the star is noticeably dimmer.

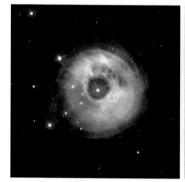 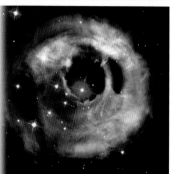

May 20, 2002   September 2, 2002

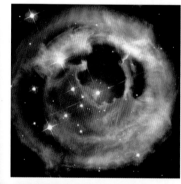 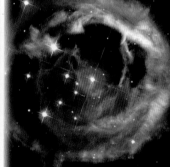

October 28, 2002   December 17, 2002

Did You Know?

Most stars have some degree of variability. The Sun has an 11-year cycle, during which its brightness varies.

These images of a variable star were taken over a period of 21 months. They show the central star getting brighter and its **light echo** (the brown halo) getting bigger.

February 8, 2004

▲ Sirius, the brightest star in the sky, is actually a binary system. This artist's impression shows Sirius A, which is twice as massive as the Sun and 25 times brighter, and Sirius B, which is a faint **white dwarf** (the small, blue star on the right).

Extrinsic Variables

The most common type of extrinsic variable is a binary star system, which is made up of two stars in orbit around each other. Although the two stars in a binary system may be of a similar magnitude, they often vary in brightness. At certain times, the dimmer star may block the brighter star and the system will appear fainter when viewed from Earth.

Multiple Star Systems

Some extrinsic variable stars are star systems with more than two stars. These are called multiple star systems. Most contain three stars, although systems with up to six stars have been discovered.

NEUTRON STARS

A neutron star is a very small and extremely dense star created when the core of a massive star collapses during a supernova explosion. It is called a neutron star because ordinary atomic matter in its core is crushed and turned into **neutrons**.

Slowing and Cooling

Neutron stars spin rapidly. Immediately after the supernova that creates it, a neutron star can rotate up to several hundred times per second, but it gradually slows down as it ages. The temperature inside a young neutron star reaches billions of degrees, but it quickly falls to about 1.8 million °F (1 million °C).

Discovering Neutron Stars

Not long after the neutron was discovered in 1932, several scientists independently proposed the existence of neutron stars. In 1967, a group of British astronomers, led by Martin Ryle and Antony Hewish, discovered the first pulsar and finally proved that neutron stars do exist.

▼ This diagram shows a cross section of a neutron star. Unlike ordinary stars, neutron stars are not made of plasma. They have a solid outer crust, and inside there is a heavy liquid consisting mostly of neutrons.

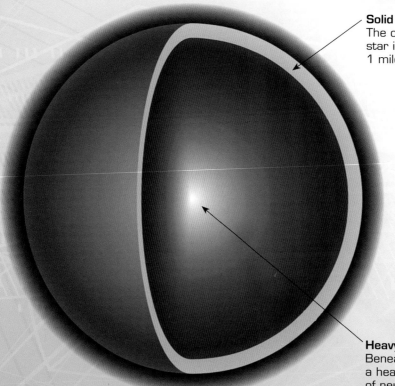

Solid crust
The crust of a neutron star is solid and about 1 mile (1.6 km) thick.

Heavy liquid interior
Beneath the crust is a heavy liquid made of neutrons.

Massive, but Small

In everyday language, we use "massive" to refer to something very large. However, scientists use massive to describe something with a lot of mass, or matter. When this matter is packed together tightly, something massive can actually be very small. This is true of neutron stars, which are typically no bigger than a city but have a mass greater than the Sun.

▼ This photograph shows the Crab Nebula, a cloud of dust and gas that remained after a supernova explosion. The white dot at the center of the nebula is the Crab Pulsar.

Pulsars

Pulsars are the most common type of neutron star. They have a particularly strong magnetic field, which funnels escaping electrons into two narrow beams at the star's magnetic poles. When viewed from Earth, these beams appear to pulse on and off as the star rotates, in the same way a lighthouse beam seems to pulse.

Did You Know?

The first pulsar, discovered in 1967, was given the name LGM1. The letters LGM stand for Little Green Men. This refers to the pulsar's regular flashes, which were originally interpreted as signals from an alien civilization.

Black Holes

While some massive stars end up as tiny neutron stars, the most massive stars of all collapse even farther and become black holes.

What Is a Black Hole?

A black hole is a region of space in which the force of gravity is so great that nothing can escape, not even light. The outer boundary of the black hole is called the event horizon. Within this boundary, no object can resist the pull of the black hole. At the center of the black hole is the singularity—a point containing the entire mass of the collapsed star. The bigger the singularity, the wider the black hole.

Falling into a Black Hole

If a spacecraft were unlucky enough to get pulled into a black hole, the force of gravity would stretch the spacecraft until it was incredibly long and thin. The remains of the craft would rapidly spiral toward the singularity, and friction would heat it up until it started emitting X-rays, a form of electromagnetic radiation. Scientists do not know what happens beyond this point.

▼ A spacecraft near a black hole is pulled in by the gravity of the singularity at its center. At a certain distance from the singularity, the speed at which the spacecraft needs to be traveling in order to escape the black hole equals the speed of light. This point is called the event horizon. Once inside the event horizon, the spacecraft cannot escape the black hole and is pulled toward the singularity.

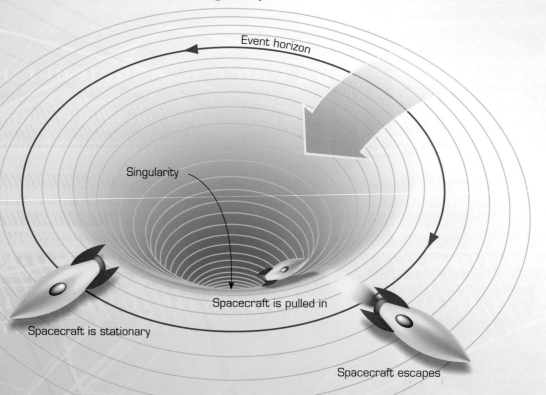

▲ In this artist's impression, the orange star is slowly being devoured by a black hole. As it is pulled toward the black hole, the star gets stretched until it is long and thin.

Finding Black Holes

It took a long time for scientists to prove that black holes exist, because they cannot be seen directly. Black holes can be detected, however, through X-rays emitted by nearby objects. For example, when one star in a binary star system collapses to form a black hole, the other star gets pulled in and starts producing X-rays. Scientists can detect these X-rays using special telescopes.

Supermassive Black Holes

Supermassive black holes contain the mass of millions or billions of stars. They can be found in the center of most galaxies, including our own.

Scientists are not sure how supermassive black holes form. They may begin as ordinary black holes, which grow gradually as they pull in matter. They may also form when clusters of stars or enormous gas clouds collapse. This happens when other forces within the clusters or clouds are not great enough to counteract the inward pull of gravity. Alternatively, supermassive black holes may have formed during the big bang.

What is a Galaxy?

A galaxy is a group of billions, or even trillions, of stars, gas, and dust held together by gravity in space. There are more than 100 billion galaxies in the universe.

Types of Galaxies

Scientists classify or group galaxies according to their shape. There are four main types of galaxies, as shown in the table below.

Edwin Hubble (1889–1953)

In 1923, American astronomer Edwin Hubble discovered that there are other galaxies besides the Milky Way. A few years later, he discovered that these galaxies are moving away from each other. In the mid-1920s, Hubble also invented a method of classifying galaxies that is still used today. It is known as the Hubble sequence.

Type of Galaxy	What It Is
	A spiral galaxy has a round bulge in its center, with a series of arms spiraling outward.
	A barred spiral galaxy is similar to a spiral galaxy, except its central bulge is in the shape of a bar, with arms coming off each end.
	An elliptical galaxy is roughly oval in shape.
	An irregular galaxy does not have an obvious shape.

Groups, Clusters, and Superclusters

Galaxies are not scattered randomly throughout space. Due to the effects of gravity, they clump together, forming groups of up to 50 galaxies. These groups of galaxies form larger structures called clusters. One cluster can contain hundreds, or even thousands, of galaxies. Clusters of galaxies form even larger clumps called superclusters.

Looking Back in Time

When we observe other galaxies, we are actually looking back in time. If a galaxy is 12 billion light-years away, light from this galaxy takes 12 billion years to reach Earth. This means we are observing the galaxy not as it is today, but as it was 12 billion years ago.

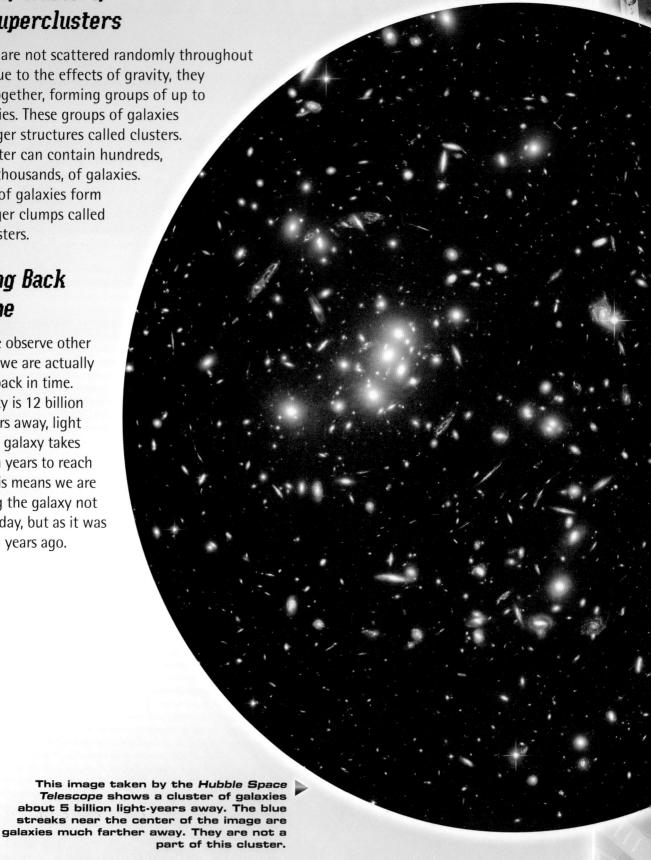

This image taken by the *Hubble Space Telescope* shows a cluster of galaxies about 5 billion light-years away. The blue streaks near the center of the image are galaxies much farther away. They are not a part of this cluster.

THE MILKY WAY

Our galaxy, the Milky Way, is a barred spiral galaxy containing more than 200 billion stars. It is 100,000 light-years in diameter.

Observing the Milky Way

Almost everything in the night sky that can be seen with the naked eye is a part of the Milky Way. Our galaxy got its name from the milky band that can be seen stretching across the sky. This band of light is our view through the galactic plane, a disc of densely packed stars that runs through the center of the galaxy.

Where Does Our Galaxy Fit in?

Our galaxy, the Milky Way, is part of a group of about 40 galaxies known as the Local Group. The Local Group is a part of the Virgo Cluster, which is in the middle of the enormous Virgo Supercluster.

▼ **An amateur astronomer gazes at the Milky Way through a telescope.**

Galaxies

▲ This illustration shows the Milky Way as it might appear when viewed from another galaxy. The Sun and our solar system are located close to the inner rim of the Orion Arm.

The Structure of the Milky Way

At the center of the Milky Way is a supermassive black hole with a mass equal to 4 million Suns. Surrounding the black hole is a central bulge of densely packed stars. The central bulge is surrounded by a flat disc made up of several arms spiraling outward. Our solar system is located on one of these arms, about halfway out from the center.

A Galaxy on the Move

The Milky Way is rotating and its stars are orbiting the supermassive black hole at the galaxy's center. Our galaxy is also moving away from other galaxies due to the continuing expansion of the universe. Scientists believe it is moving at almost 1.2 million miles (2 million km) per hour.

Did You Know?

The single largest part of the Milky Way is its mysterious halo of **dark matter**—a roughly circular area that extends for hundreds of thousands of light-years beyond the visible part of the galaxy. Astronomers estimate that about 95 percent of the galaxy's mass is made up of dark matter.

Active Galaxies

An active galaxy is one that produces huge amounts of energy, usually from a tiny area at its center called the active galactic **nucleus** (AGN). Scientists believe that an active galaxy's power source is a supermassive black hole with an exceptionally large mass.

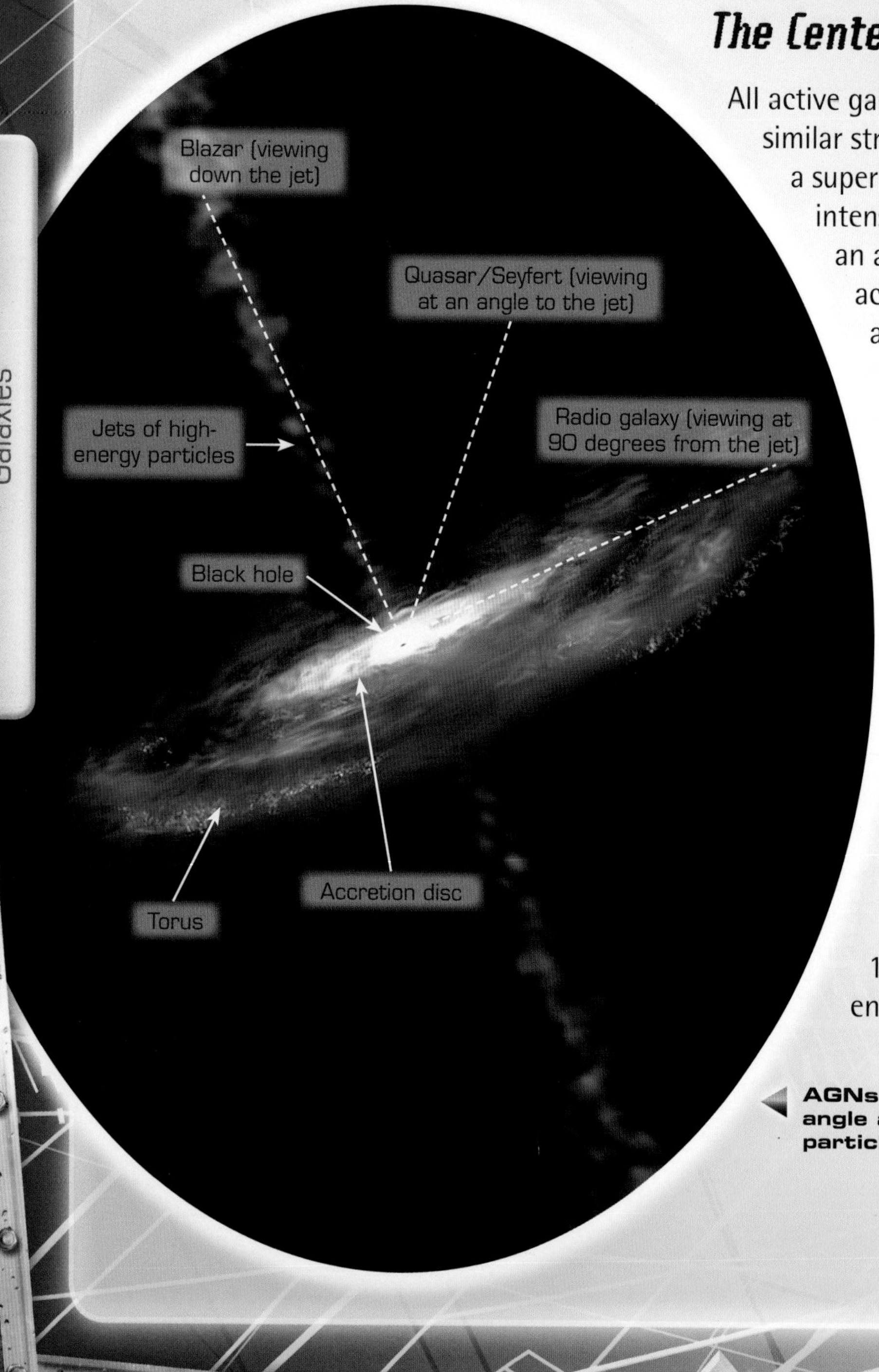

The Center of an Active Galaxy

All active galaxies have an AGN with a similar structure. Spinning rapidly around a supermassive black hole is a flat and intensely hot disc of gas known as an accretion disc. Surrounding the accretion disc is a torus, which is a donut-shaped ring of gas and dust. There are four main types of AGNs: quasars, radio galaxies, Seyferts, and blazars. AGNs are characterized by the angle at which we view them from Earth.

Quasars

A quasar is a very distant galaxy with an AGN. Most quasars are more than 10 billion light-years away, near the edge of the observable universe. Even though they are so distant, they look starlike from Earth, because they emit about 100 times more energy than the entire Milky Way.

AGNs are classified according to the angle at which the jets of high-energy particles are viewed from Earth.

Labels on diagram:
- Blazar (viewing down the jet)
- Quasar/Seyfert (viewing at an angle to the jet)
- Jets of high-energy particles
- Radio galaxy (viewing at 90 degrees from the jet)
- Black hole
- Accretion disc
- Torus

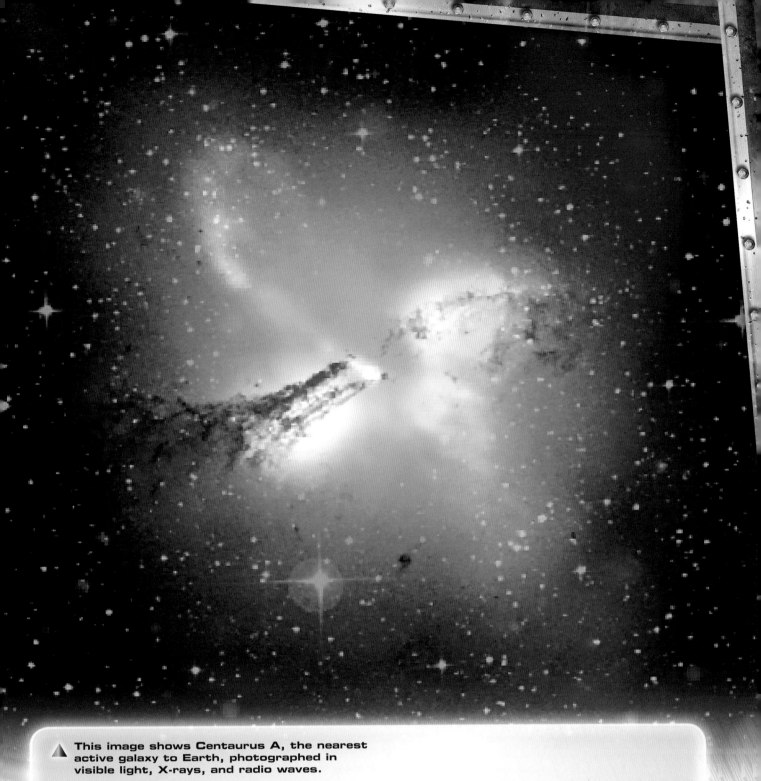

▲ This image shows Centaurus A, the nearest active galaxy to Earth, photographed in visible light, X-rays, and radio waves.

Radio Galaxies

Radio galaxies have AGNs, which emit a lot of radio waves. When this type of AGN is viewed sideways, jets of high-energy particles can be seen clearly, extending out on either side. Radio galaxies are usually elliptical galaxies. Scientists do not yet know the reason for this.

Seyferts and Blazars

Blazars have AGNs that are similar to those of radio galaxies. However, from Earth the torus of a blazar is seen face-forward. This means its separate jets are not visible. Seyferts are similar to quasars but appear fainter. This is probably because the black hole at the center of a Seyfert is not as massive.

THE EXPANDING UNIVERSE

For thousands of years, scientists believed the universe was a fixed size and was more or less unchanging. When it was discovered in the 1920s that the universe was, in fact, expanding, the scientific community was shocked.

Evidence of Expansion

In the 1920s, American astronomer Edwin Hubble was studying spiral nebulae. He calculated the distances to these nebulae and found that they are far too distant to be a part of the Milky Way. He reasoned that they must be separate galaxies. Hubble also discovered that these galaxies are moving away from the Milky Way at incredible speeds.

▶ This diagram explains the concept of the expanding universe.

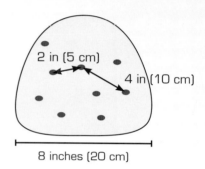

Imagine the universe is a lump of dough and the galaxies are raisins in the dough.

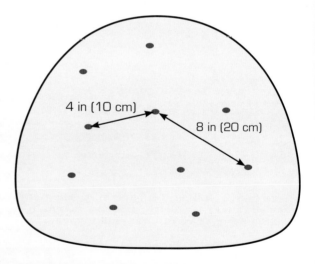

When the dough is baked, it expands, but the raisins maintain their size and shape. The expansion of the dough, however, results in the raisins being farther away from each other.

Red Shift

When a galaxy moves away from us, the waves of electromagnetic radiation that it emits are stretched. This increase in wavelengths is called red shift, because visible light becomes redder when its wavelength increases. By measuring the degree to which a galaxy's light has been red shifted, it is possible to calculate the speed at which it is moving away from us.

▲ This artist's impression shows the *Wilkinson Microwave Anisotropy Probe (WMAP)*, launched in 2001. Information collected by *WMAP* has helped scientists make more accurate estimates of the age of the universe, the speed of its expansion, and the types of matter it is composed of.

Hubble's Law

Hubble discovered that the speed at which other galaxies are moving away from the Milky Way depends on how far away they are, with the most distant galaxies moving away from us most quickly. The exact relationship between speed and distance is described in an equation known as Hubble's law. It allows scientists to calculate how far away any galaxy is by measuring how fast it is moving away.

Uneven Expansion

The universe is not expanding evenly in every direction. The Milky Way and thousands of other neighboring galaxies are speeding towards the Virgo Supercluster at 373 miles (600 km) per second. Some scientists believe this motion is most likely caused by the gravitational pull of a single object. This theoretical object, called the Great Attractor, would need to have a mass equal to 50 trillion Suns.

THE BIRTH OF A THEORY

In 1931, Georges Lemaître, a Belgian physicist and Roman Catholic priest, put forward a theory to explain the expansion of the universe. He suggested that the universe started off very small and was blown apart by an enormous explosion. This theory became known as the big bang theory.

How Did the Universe Begin?

The universe came into existence about 13.7 billion years ago. It began as an almost infinitely hot and dense speck, which exploded with unimaginable force to create space, time, and matter. The energy produced by the explosion was so great that matter began to appear spontaneously. The universe expanded very quickly. Less than one second after the big bang, the universe had grown from smaller than an atom to the size of a galaxy.

Did You Know?

All matter that exists in the universe today was created in the first few seconds of the big bang, when pure energy turned into different particles. It was only later that these particles came together to form atoms.

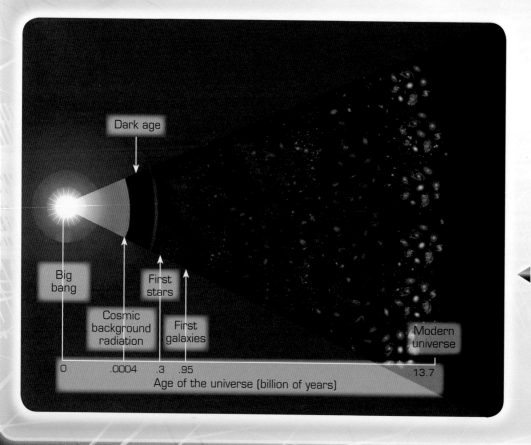

This diagram shows the expansion of the universe from the big bang to the present day. The energy released by the big bang was so great that it is still driving the expansion of the universe today.

Evidence for the Big Bang Theory

When the big bang theory was first put forward, scientists agreed that if it was correct, then it should be possible to detect radiation from the explosion as a faint background glow, mainly consisting of microwaves and filling the entire sky. Scientists named this glow the **cosmic background radiation** (CBR). When the CBR was eventually discovered, it provided one of the most compelling pieces of evidence for the big bang theory.

How Was the CBR Discovered?

When American physicist Arno Penzias and astronomer Robert Wilson were testing a new and sensitive **radio telescope** in 1965, they heard a faint noise that seemed to be coming from all over the sky. They tried everything to get rid of it, but nothing worked. They had accidentally discovered the CBR.

▼ **Arno Penzias (left) and Robert Wilson (right) stand in front of the radio telescope with which they discovered the CBR.**

Tune into the Big Bang

Televisions work by decoding television signals, which are electromagnetic radiation in the form of radio waves and microwaves. If a television is tuned between stations, white noise—a snowy picture and a hissing sound—will be seen and heard. About one percent of this is the CBR.

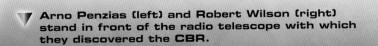

LOOKING FOR LIFE

Earth is the only planet in our solar system capable of supporting life, but what about planets in other solar systems? Given the vastness of the universe and the fact that more and more stars with planetary systems are being discovered, some scientists believe it is highly likely that life of some sort has evolved elsewhere.

Calculating the Possibility of Life Existing Elsewhere

The Drake equation is a way of estimating the number of alien civilizations in the galaxy that we could communicate with. This depends on a number of factors, the values of which are not known and have to be guessed. Some scientists have used the Drake equation to suggest that there could be millions of civilizations in the Milky Way, while others believe that even life on Earth was a fluke!

▼ Alien life forms, such as this whale-like animal, may have evolved outside our solar system. Scientists are still searching for them.

The Drake Equation

The Drake equation states that the number of alien civilizations in the Milky Way that we could communicate with is equal to:
- the average rate of star formation in the galaxy
- multiplied by the fraction of those stars that have planets
- multiplied by the average number of planets in each star's **habitable zone**
- multiplied by the fraction of those planets on which life develops
- multiplied by the fraction of those planets on which intelligent life evolves
- multiplied by the fraction of civilizations that develop a technology for sending signals into space
- multiplied by the length of time such civilizations send signals for

Frank Drake (1930–)

Frank Drake is an American astronomer and astrophysicist. He was one of the organizers of the first Search for ExtraTerrestrial Intelligence (SETI) conference in 1961. It was there that he proposed his famous Drake equation, which he hoped would encourage other scientists to take up the search for **extraterrestrial** life.

This artist's impression shows a Jupiter-like exoplanet orbiting a star called Epsilon Eridani. Though the gas giant planet is uninhabitable, any moons orbiting it might have conditions suitable for life.

The SETI Program

SETI stands for the Search for ExtraTerrestrial Intelligence. It is a collective name for various projects carried out around the world. Radio telescopes are used to scan the sky for transmissions from civilizations on distant planets, and computers analyze the collected data.

The first SETI experiment was conducted in 1960 by Frank Drake. He focused a radio telescope on two of the nearest Sun-like stars, Tau Ceti and Epsilon Eridani. He found nothing of interest.

Nowadays, millions of ordinary people around the world belong to the SETI@home project. When they are not using their computers, the SETI program uses the computers to process data.

Did You Know?

Scientists do not just listen for alien messages. They also send out messages. In 1974, a message was sent from the Arecibo Observatory in Puerto Rico, aimed at the M13 star cluster, 25,000 light-years away. If a reply is sent, we will have to wait 50,000 years to receive it.

Searching for Exoplanets

One aspect of the search for extraterrestrial life is the search for planets outside our solar system. These are called exoplanets. Scientists have found more than 400 exoplanets orbiting nearby stars. Most exoplanets discovered so far are **gas giants**, similar to Jupiter. Scientists believe improvements in detection methods will eventually reveal smaller, rocky planets similar to Earth. It is on such planets that life is most likely to have evolved.

THE FUTURE OF THE UNIVERSE

Evidence suggests that the universe probably started with a big bang, but how will it end? The eventual fate of the universe will depend on the amount of matter (mass) it contains. This is because gravity is the force that shapes the universe, and gravity is determined by mass. Three possible scenarios have been put forward to explain how the universe might end.

The Big Freeze

If the total mass of the universe is below a certain level, gravity will not be strong enough to resist the expansion started by the big bang, and the universe will continue expanding forever. As everything moves farther and farther apart, the influence of gravity will decrease. Stars will not be able to form, and when they die, new stars will not replace them. The universe will become a very cold and empty place, containing just a few, distantly spaced particles.

The Big Crunch

If the total mass of the universe is great enough, the force of gravity will gradually slow down its expansion, and at some point, the universe will begin to shrink. As galaxies become more tightly packed, the effect of gravity will increase, and the universe will shrink at an ever faster rate. Eventually, it will end the opposite of a big bang—the big crunch.

The Big Rip

The universe is expanding at an increasing rate. Scientists believe this is caused by a mysterious force called dark energy. Dark energy is the opposite of gravity. Instead of pulling matter in, it pushes matter apart. It is possible that dark energy could eventually tear apart all matter in the universe in a violent event called the big rip.

▼ The big rip is one possible way that the universe might end. In this artist's impression, dark energy is pushing the planets and galaxies away from each other.

GLOSSARY

atmosphere
the layer of gases surrounding a planet, moon, or star

atoms
the building blocks of matter, consisting of protons, electrons, and neutrons

big bang
the large explosion believed to have created the universe

black hole
a region of space where gravity is so powerful that nothing can escape, not even light

constellation
an area of the sky containing a group of stars that form a recognizable pattern

cosmic background radiation
a form of electromagnetic radiation that fills the universe but is not associated with any space object

dark matter
matter that is not visible through telescopes, because it does not emit or reflect light or other forms of electromagnetic radiation

electromagnetic radiation
waves of energy created by electric and magnetic fields

elements
pure chemical substances, which cannot be reduced to a simpler form

extraterrestrial
coming from outside Earth, or a being not from Earth

galaxy
a large system of stars, gas, and dust, held together by gravity

gas giants
large planets made mostly of gas and with a metal or rock core

gravity
the strong force that pulls one object toward another

habitable zone
the area in a solar system or galaxy in which conditions are favorable for life

light echo
light from a bright object, such as a star, which is reflected off another object, such as a collection of gas and dust nearby

light-years
each light-year is the distance that light travels in one year, which is about 5.9 trillion miles (9.5 trillion km)

neutrons
particles with no electric charge, found in the nucleus of atoms

neutron star
a very hot, small, and dense star, left behind by a supernova

nuclear fusion
the process in which the nuclei of two or more atoms fuse together to form a single atom with a heavier nucleus, releasing huge amounts of energy

nucleus
in astronomy, nucleus refers to the central region of a galaxy; in physics, nucleus refers to the central part of an atom, where protons and neutrons are found

orbited
followed a curved path around a more massive object while held in place by gravity; the path taken by the orbiting object is its orbit

planetary nebulae
glowing shells of gas formed by some types of stars when they die

plasma
super-heated and electrically charged gas

radio telescope
a telescope that detects radio waves rather than visible light

red giants
huge stars in a late stage of their evolution

solar system
the Sun and everything in orbit around it, including the planets

white dwarfs
small, dense stars that are very faint and in the final stages of their evolution

INDEX

A
active galactic nucleus (AGN) 22, 23
Arecibo Observatory 29
Aristarchus 5
atoms 6–7, 14, 26, 31

B
big bang, the 5, 6–7, 17, 26–27, 30, 31
big crunch, the 30
big freeze, the 30
big rip, the 30
binary systems 13
black holes 11, 16–17, 21, 22, 23, 31
blazars 22, 23

C
constellations 31
Copernicus, Nicolaus 5
cosmic background radiation 27, 31

D
dark energy 30
dark matter 21, 31
Drake equation, the 28
Drake, Frank 28, 29

E
electromagnetic radiation 8, 16, 24, 27, 31
elements 8, 31
exoplanet 29
extrinsic variables 12, 13

G
galactic nucleus 22, 23
galaxies 5, 6, 9, 18–23, 24, 26, 28, 30, 31
galaxy cluster 19
gas giants 29, 31
geocentric model 5
giant star 9
gravity 8, 10, 16, 18, 19, 30, 31
Great Attractor, the 25

H
habitable zone 28, 31
heliocentrism 5
Hewish, Antony 14
Hubble, Edwin 18, 24, 25
Hubble sequence 18
Hubble's law 25

I
intrinsic variables 12

K
Kepler, Johannes 11
Kepler's Supernova 11

L
Lemaître, Georges 26
LGM1 15
light echo 12, 31
light-years 9, 10, 19, 20, 21, 22, 29, 31
Local Group 20

M
magnitude 9, 13
main sequence 9, 11
Milky Way, the 5, 11, 18, 20–21, 22, 24, 25, 28
multiple star system 13

N
nebula 10, 24
neutrons 14, 31
neutron star 11, 14–15, 16, 31
nuclear fusion 8, 9, 10, 31

O
Orion Nebula 10

P
Penzias, Arno 27
planetary nebulae 11, 31
plasma 8, 14, 31
pulsars 14, 15

Q
quasars 22, 23

R
radio galaxies 22, 23
radio telescopes 27, 29, 31
red giants 11, 31
red shift 24
Ryle, Martin 14

S
SETI (Search for Extraterrestrial Life) 28, 29
Seyferts 22, 23
solar system 4, 5, 21, 28, 29, 31
solar wind 10
star clusters 29
stars 5, 6–7, 8–17, 18, 20, 21, 22, 28, 29, 30
Sun, the 5, 6–7, 8, 9, 11, 12, 13, 21
supercluster 19
supergiant 9, 11
supermassive black hole 17, 21, 22
supernova 11, 14

V
Virgo Cluster 20
Virgo Supercluster 20, 25

W
white dwarfs 11, 13, 31
Wilson, Robert 27